MW00559055

Niagara Falls

Survivor of the Ice Age

The Natural History of the Niagara River and its Gorge

---

by Dr. Paul A. Young

RPSS – Rock, Paper, Safety Scissors Publishing, 429 Englewood Avenue; Kenmore, New York 14223
publisher@rockpapersafetyscissors.com • rpsspublishing.com

978-0-9977996-7-5 Niagara Falls Survivor of the Ice Age – Case Bound

10 9 8 7 6 5 4 3 2 1

Printed in the United States of America

Contents

Introduction

One can only imagine the reactions of the paleonatives who first arrived in the Niagara Region, hearing from afar the thunderous roar and the great plume of smoke-like mist rising above the trees of the North American Hardwood Forest.

As they got closer, the sounds became deafening, the mist intensified in height and volume, and they could begin to feel the pelting spray and see the hues of the rainbow visible above the landscape. Then, stepping into an opening of the foliage, they were standing at the brink of one of North America's greatest natural formations–a breathtaking and wondrous sight of an array of falls with the power of hundreds of thousands of gallons of water per second plummeting over the rock ledges in a violent, bubbling, and greenish torrent into a swirling pool almost two hundred feet below!

Today, though the landscape has changed, 21st century visitors can get an enhanced view of this natural spectacle and can intensify the Native American experience. Looking out from the deck of the Maid-of-the-Mist boat as it battles its way against the churning waters to the base of the Horseshoe Falls, the views and sensations are spectacular. For an aerial panorama, a helicopter flight can provide a history of the river and falls that has moved from its origin 36 miles to the north and a timespan of some 12,000 years. From both the Cave of the Winds on the American and Canadian sides of the gorge, visitors can witness the sights and sounds of the cataracts from behind the Horseshoe Falls and from the rock debris at the base of the American Falls. The tower at the American Falls provides an opportunity to walk the talus at the base of the falls and in winter, view the magical snowy and icy formations.

Or we can also share in the compelling attraction that is the falls by just spending time near its brink and contemplating our life and place in the natural universe!

The Present Great Lakes Drainage Basin

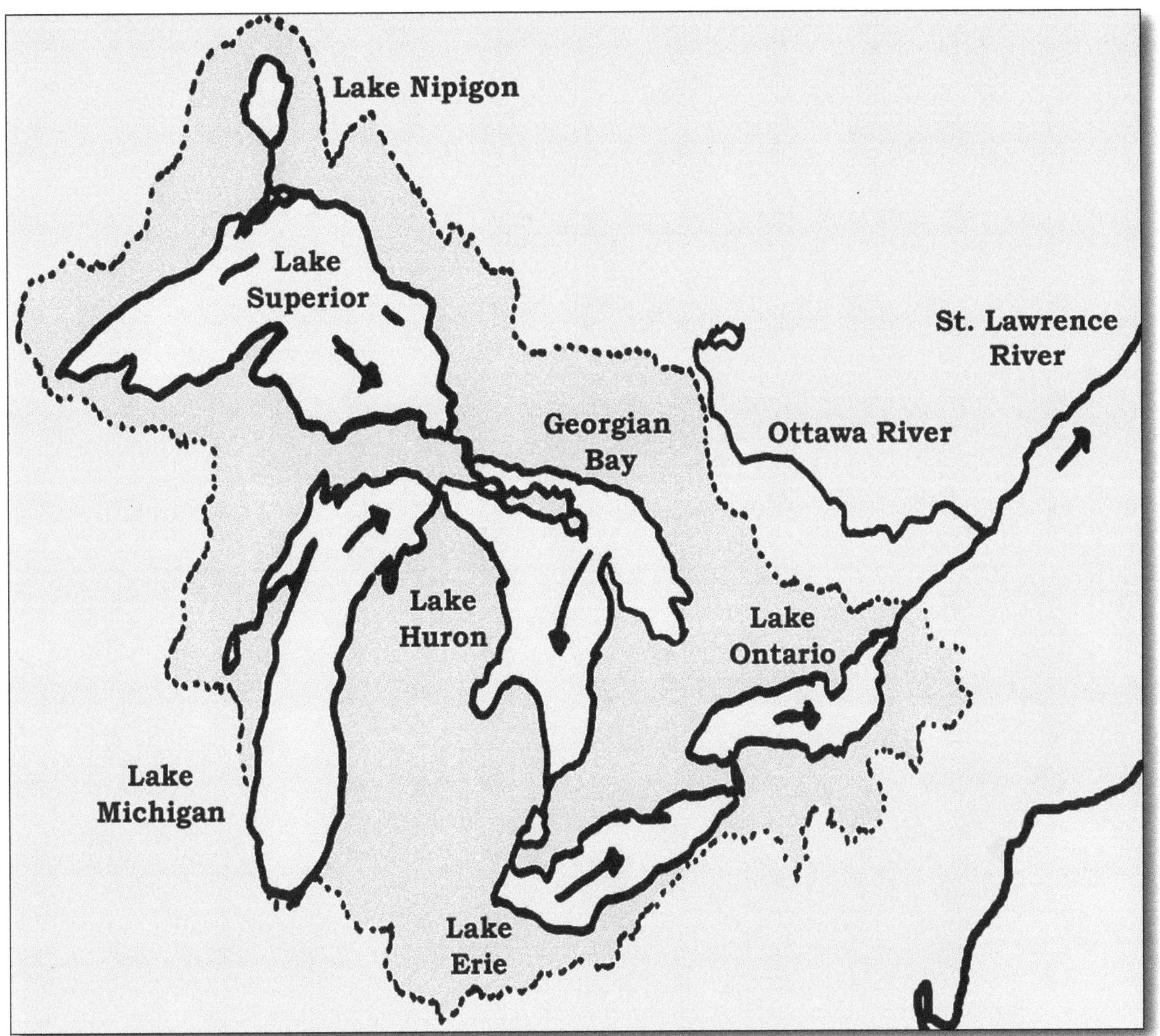

***Diagram 1**: The Present Great Lakes Drainage Basin*

The view above of the five Great Lakes and their present drainage pattern provides both a beginning and an ending point in our story. As can be seen in **Diagram 1**, *The Present Great Lakes Drainage System*, the waters from four of the five Great Lakes currently cascade over the falls at Niagara through the gorge to Lake Ontario, then flow through the St. Lawrence River to the Atlantic Ocean. Any water outside this drainage basin flows to other drainage systems such as the Mississippi River basin.

Overview of the Present Geographic Regions Near Lake Ontario

***Diagram 2**: Geographic divisions near Lake Ontario*

***Diagram 3**: Tilt of Rock Structures in the Lake Ontario region*

Before exploring the phenomena of the Niagara Gorge, a brief overview of the topography surrounding the region of Niagara Falls would be relevant. This reveals three prominent escarpments. An escarpment or cuesta is a landform that slopes gently back from a steep cliff. As shown in the diagrams below, the Niagara Escarpment is at the southern edge of the Lake Erie Plain and is the cause and continuance of Niagara Falls. Roughly parallel and to the south about 25 miles is the Onondaga Escarpment with a separate set of local falls and 35 miles farther south is the Portage Escarpment, the beginning of the foothills of the Allegheny Plateau. Each formation dips to the south at ½ of a degree or 50 feet per mile.

Any stream that originates above the escarpment and flows toward Lake Ontario or the Georgian Bay results in a waterfall. In Canada, the Bruce Trail follows the Niagara Escarpment from Niagara Falls to Tobermory and contains dozens of waterfalls. In the United States, the Niagara Escarpment extends only a little more than a hundred miles and other than Niagara Falls, contains only four waterfalls with reduced flow due glacial action.

The Niagara Escarpment is a UNESCO World Biosphere Site because of its geological and biological venues, formations and ecosystems. It is about 450 miles in length with its highest elevation about 1625 feet above sea level and is the most prominent escarpment of the three formed in the bedrock of the Great Lakes Basin in the United States and Canada. It runs mainly in an east/west direction from New York, Ontario, Michigan, Wisconsin, and Illinois. Some 400 to 450 million years ago, the capstone was being formed in a vast shallow inland tropical sea above the compression into rock of layers of clays, muds, sand and shell sediments. Later, these rocks were tilted at an angle, and viewed from space, resemble the edge of a saucer-shaped shard which represents the shoreline of that tropical sea. Recession of Niagara Falls is the result of differential erosion of these rocks of various hardnesses. The harder caprock of the Lockport Dolostone provides resistance to weathering and erosion while the weaker, more easily eroded shales and sandstone-shale mixtures are undercut. This process of undercutting and collapse of the overlying capstone has enabled Niagara Falls to move back along the Niagara River some 7 miles from its origin at Lewiston/Queenston in the past 12,000 years.

Historically, the Niagara Escarpment is shared by the United States and Canada and is important because of the strategic location of the portage on the east side of the Niagara River gorge around Devil's Hole. This enabled passage around Niagara Falls and provided access to the resources of the Upper Great Lakes and the water connections of the Ohio and Mississippi River systems. Various wars between First Nations peoples, the French, English and Americans have changed control over this access point over the centuries and emphasize its importance and impact on world events.

The Course of the Present Niagara River

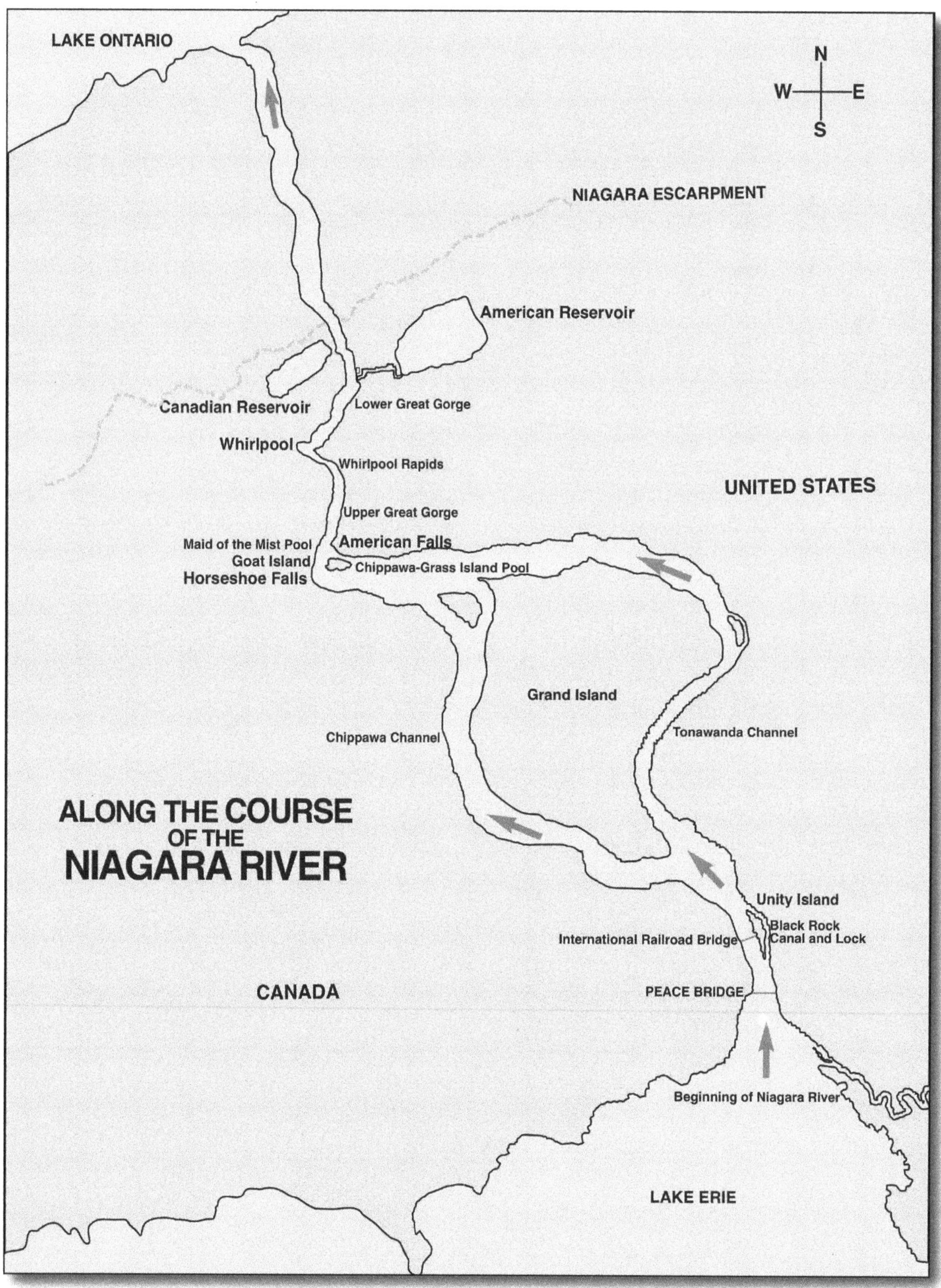

The present 36 mile long Niagara River starts at Buffalo, New York. By the time it empties to the north into Lake Ontario, it has dropped 326 feet. Geographically, it is not a river at all, but a strait. A strait is a narrow body of water connecting two larger bodies of water, in this case, Lake Erie and Lake Ontario.

Its flow, elevation, and width changes over its length depending on natural formations and man-made structures. From where the river starts at Buffalo some 569 feet above sea level, the elevation drops 5 feet and immediately narrows to a width of approximately 1500 feet as it reaches a rock ledge which naturally controls its outflow and speed. In addition, the abutments and their protective masses of rock under the Peace Bridge and International Railway Bridge further restrict the channel. Between the two bridges for the next 4 miles, the elevation drops another 5 feet. In this area, Squaw (Unity) Island and the Bird Island Pier form a parallel channel that serves as the Black Rock Canal and Lock enabling commercial vessels to bypass the shallow and fast moving water at the head of the river. The lock compensates for the 5 foot drop and is the terminus of the western end of the Erie Canal.

For the next 18 miles, the river falls only 4 feet and widens to about 2400 feet, then separates into two channels, the Chippewa and Tonawanda, surrounding Grand Island, New York. Downstream it rejoins, becomes shallow, and broadens to 7000 feet, forming the Chippewa-Grass Island Pool. Here, both the United States and Canada divert water for power generation. A mile farther and the river drops another 55 feet, resulting in the cascades and rapids just above Niagara Falls. At this point, the river again divides into two channels around Goat Island before it becomes the American Falls to the right and the Horseshoe Falls on the left, plunging 160-180 feet into the Maid-of-the-Mist Pool. The smaller American Falls channel is further divided by a number of small islands, one of which is Luna Island, located on the brink of the Falls and divides it into two separate cataracts, the Bridal Veil Falls and the much larger American Falls.

Below the falls, the Maid-of-the-Mist pool is relatively flat, dropping only about 1.5 feet, and providing a large navigable area for tourist vessels. The river then drops another 96 feet in the next 6 miles before entering the narrow, shallow channel of the Whirlpool Rapids. Within the space of a mile, the river drops 50 feet, providing a wild and turbulent sight.The rapids terminate at the Whirlpool, a circular basin formed by the excavation of the glacial deposits of a previous river bed predating the Niagara River. It is here that the channel makes a 90 degree turn and continues another 4 miles through the Lower Rapids.

Leaving the gorge at Queenston-Lewiston, which was the site of the beginning of the falls some 12,000 years ago at the Niagara Escarpment, the river widens out and drops 4 feet in its final run of 7 miles to Lake Ontario.

Glaciers and their Moraines

***Photograph 1**: Wordie Glacier and Terminal Moraine*

During the last Ice Age, the whole area surrounding the Great Lakes was covered by thousands of feet of ice as the Wisconsin glacier, the last of at least four glaciers to blanket the area, made its way from the north to the south.

Glaciers exert tremendous pressure on the underlying rock structures. They gouge out new landscapes while depositing a variety of different landforms.

As the climate changed once more, the glacier started to melt and began its retreat back up toward the north, leaving the landscape dramatically changed and forcing river channels to take new courses.

Depicted in ***Photograph 1**: Wordie Glacier and Terminal Moraine* is a terminal moraine which indicates the farthest extent of this glacier. Moraines are unassorted glacial till that forms when the retreat of a glacier is steady or slow, resulting in an irregular depositional pattern. Recessional moraines form between the terminal moraine and the glacial source and are the result of sporadic and retreating glacier melting. An outwash plain is shown beyond the terminal moraine as an extensive low relief landscape.

Photograph 2: *A Terminal Moraine Composed of Unassorted Glacial Till*

Diagram 4: *The Distribution of Moraines in the Niagara Area*

The Five Sections of the Niagara Gorge

The present Niagara Gorge is divided into five distinct sections shown on ***Diagram 5****: The Five Sections of the Niagara Gorge*. Each section formed as a result of the amount of water going over the falls at the time and the height from which it drops.

The combination of more water and more height gives Niagara Falls its greatest erosive power.

During the journey from its beginnings over the Niagara Escarpment at Lewiston/ Queenston and eroding back to its present site at Niagara Falls, seven miles to the south, each one of these two factors has been in different proportion, creating these five sections. It is this pattern of widening, narrowing, and deepening, together with the effects of the last glacier that provides the individual explanation of these causes and differences and provides the organization for this book.

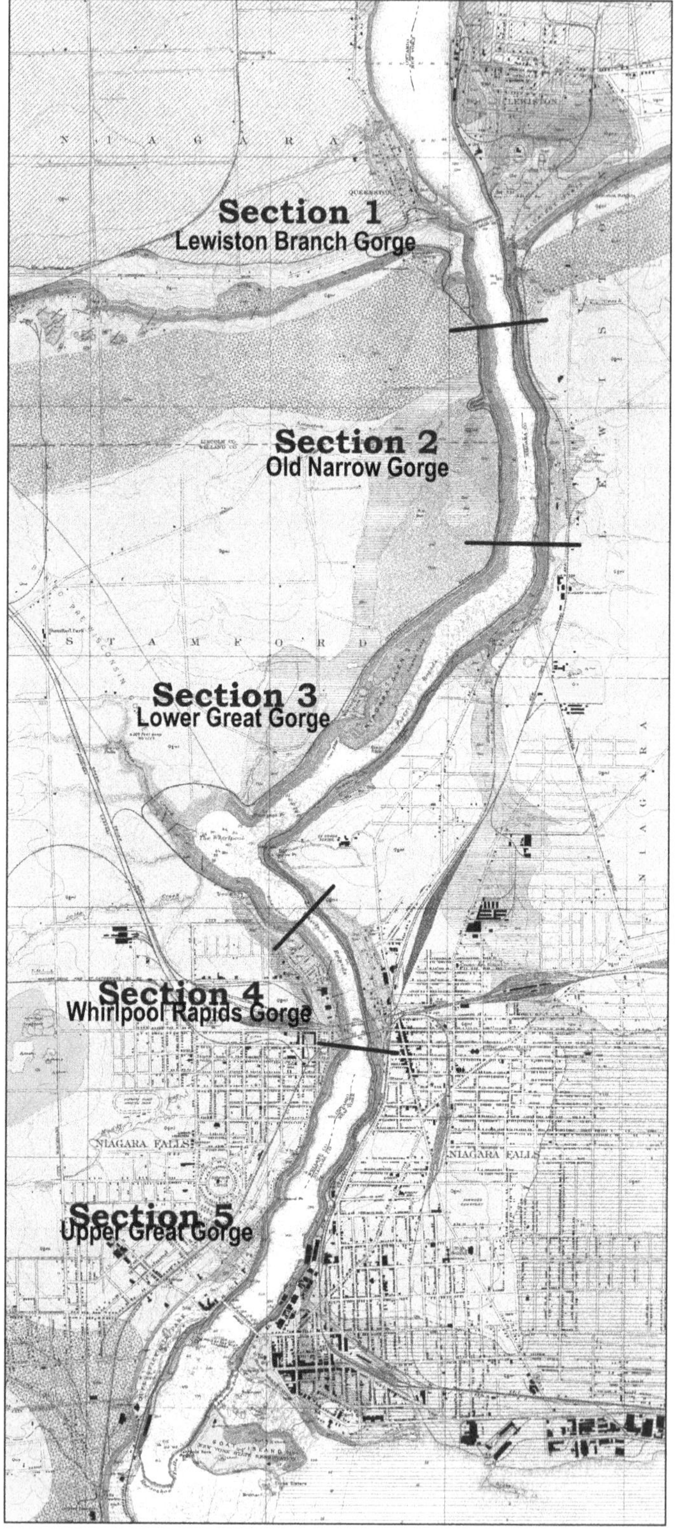

Diagram 5*: The Five Sections of the Niagara Gorge*

Section 1: The Lewiston Branch Gorge

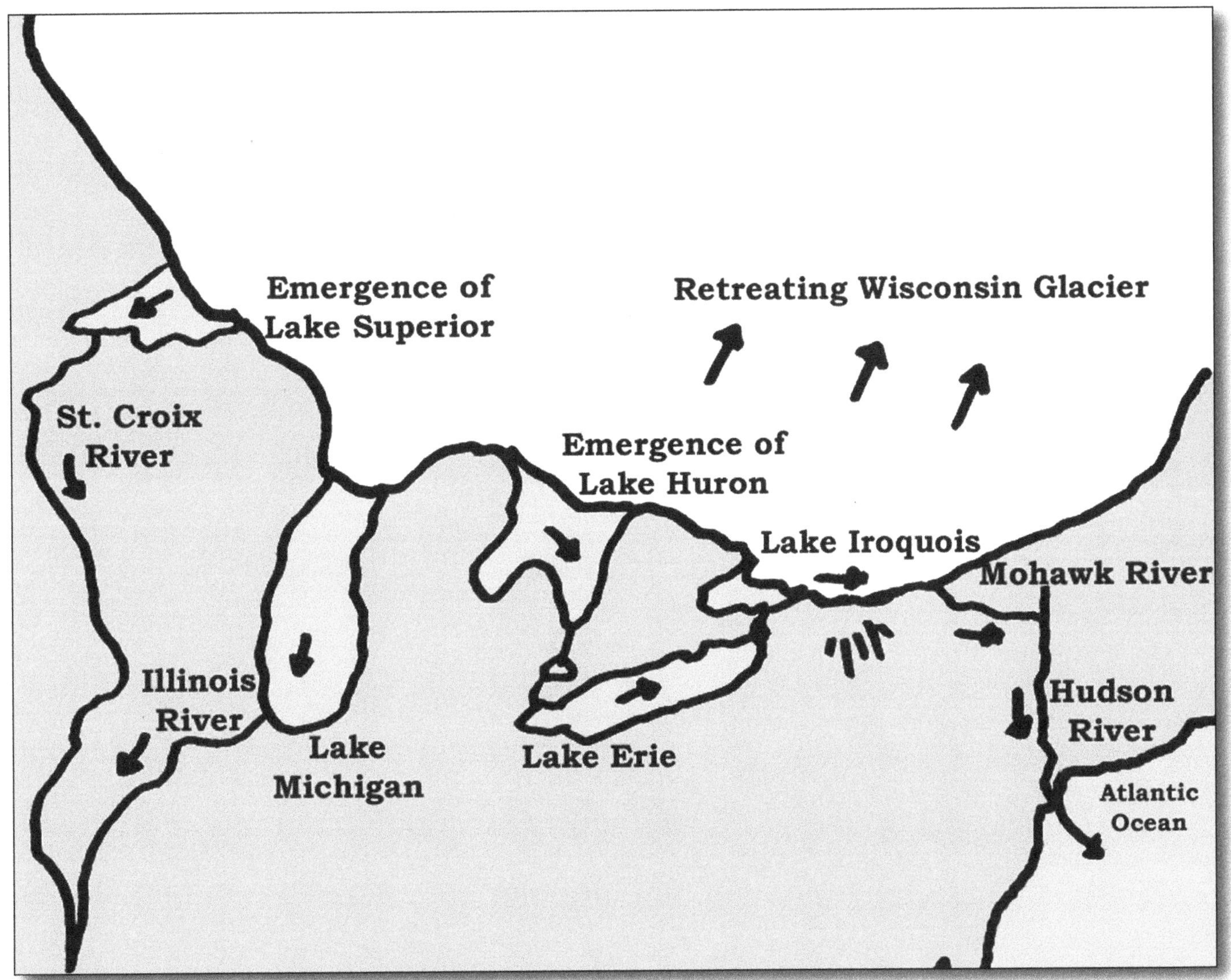

Diagram 6: *The Formation of the Great Lakes*

Referring to ***Diagram 6*** above, *The Formation of the Great Lakes*, depicts the beginnings of the Great Lakes, with the drainage patterns of each lake significantly differing from the present one. The newly forming Lake Superior basin was draining south through the St. Croix River to the Mississippi River and on to the Gulf of Mexico. The Lake Michigan basin water was flowing south via the Illinois River to the Mississippi, and also east along its low northern edge, which would eventually become Lake Huron. In turn, Huron's waters would enter the newly formed Lake St. Clair into Lake Erie, over the falls at Lewiston/ Queenston, into glacial Lake Iroquois, the predecessor to Lake Ontario, through the lowlands of New York State drained by the Mohawk River, to the Hudson River and on to the Atlantic Ocean.

Photograph 3: *Rock Structure at the Beginning of Niagara Falls at Lewiston/Queenston*

Three distinct rock layers can be traced along the gorge as they dip to the south changing their appearance, thickness, and height above the water level. The top layer is the Lockport dolostone, here a hard resistant caprock of limestone over the easily eroded layers beneath, causing the recession of the Falls. The second resistant layer is the Irondequoit layer and the third is the Whirlpool sandstone. Each is clearly visible as we follow them back the seven miles to Niagara Falls. See ***Photograph 3:*** *Rock Structure at the Beginning of Niagara Falls at Lewiston/Queenston.* Also, by this time, the Finger Lakes in central New York State had already been created by the damming action of moraine left by the glacier at the northern end of some of the valleys that had a north/south orientation to the retreating ice front.

Section 2: The Old Narrow Gorge

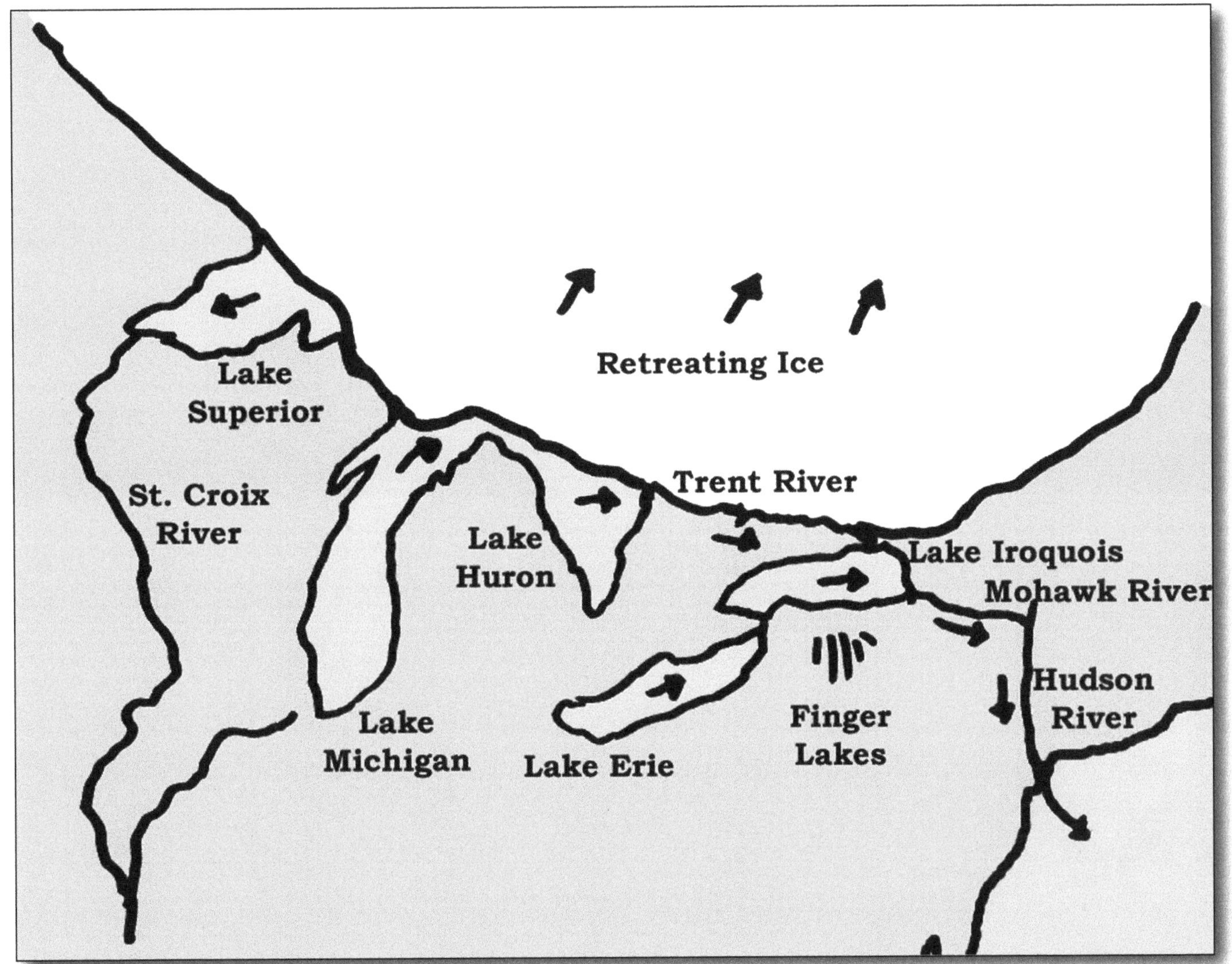

Diagram 7: *Drainage of the Upper Great Lakes through the Trent River Basin, the Narrowing of the Gorge*

As the glacier continued to melt, the drainage pattern shifted once more in response to new lowlands being exposed, as shown in ***Diagram 7:*** *Drainage of the Upper Great Lakes through the Trent River Basin, the Narrowing of the Gorge.* The Trent River basin emerged from under the ice, diverting water from the St. Clair region, spilling into Lake Iroquois, then entering the Atlantic Ocean via the Mohawk and Hudson Rivers. Lake Michigan no longer drained to the south, but Lake Superior continued to do so. At this time, only the water from Lake Erie flowed over the falls and because of this reduced water level and, therefore, diminished erosive power, the width of the gorge narrowed considerably, as can be seen in the following ***Photograph 4:*** *The Narrowing of the Gorge.*

Photograph 4: *The Narrowing of the Gorge*

In ***Photograph 4:*** *The Narrowing of the Gorge* note the thickening of the capstone Lockport dolostone at the point where the Lockport/Queenston Bridge crosses the gorge at this narrow area. To the left is the Canadian power project.

Section 3: Lower Great Gorge

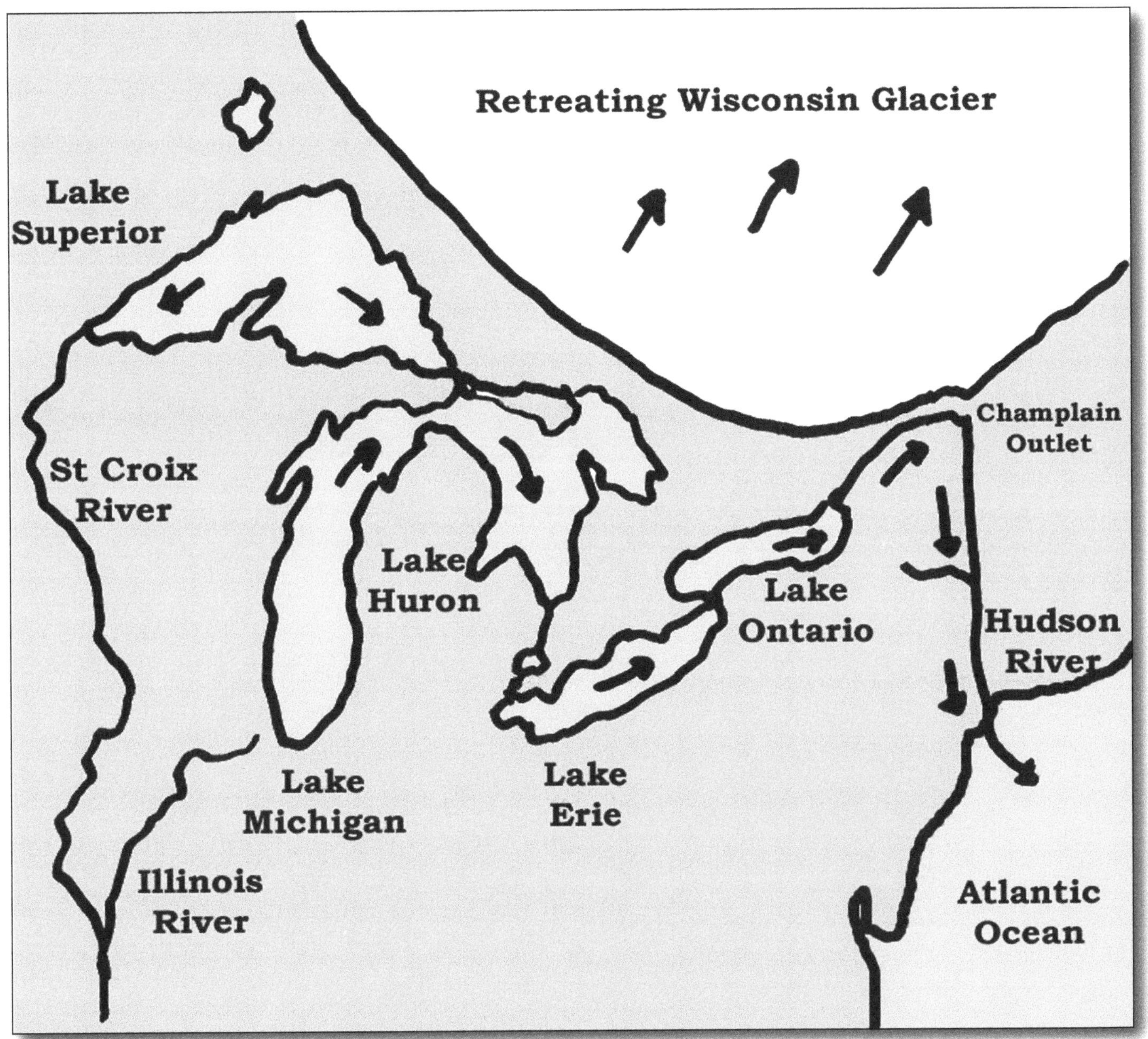

***Diagram 8**: The Opening of the Champlain Outlet*

The next events shaped the gorge into the third noticeable division, as shown in ***Diagram 8**: The Opening of the Champlain Outlet.* As the St. Lawrence River channel was gradually uncovered, it left the Mohawk valley at a considerably higher elevation. Thus the water from the four Great Lakes was, for the most part, able to flow over the falls, again widening the gorge. As the water left Lake Iroquois, it made its way through the St. Lawrence valley as far as the ice front would allow, turned south through the lowlands of the Champlain Outlet (later to become the Richelieu River and Lake Champlain), to the Hudson River, and as before, into the Atlantic Ocean. Lake Iroquois' level began to drop and resulted in some interesting geological features in this newly widening portion of the gorge.

Lake Tonawanda

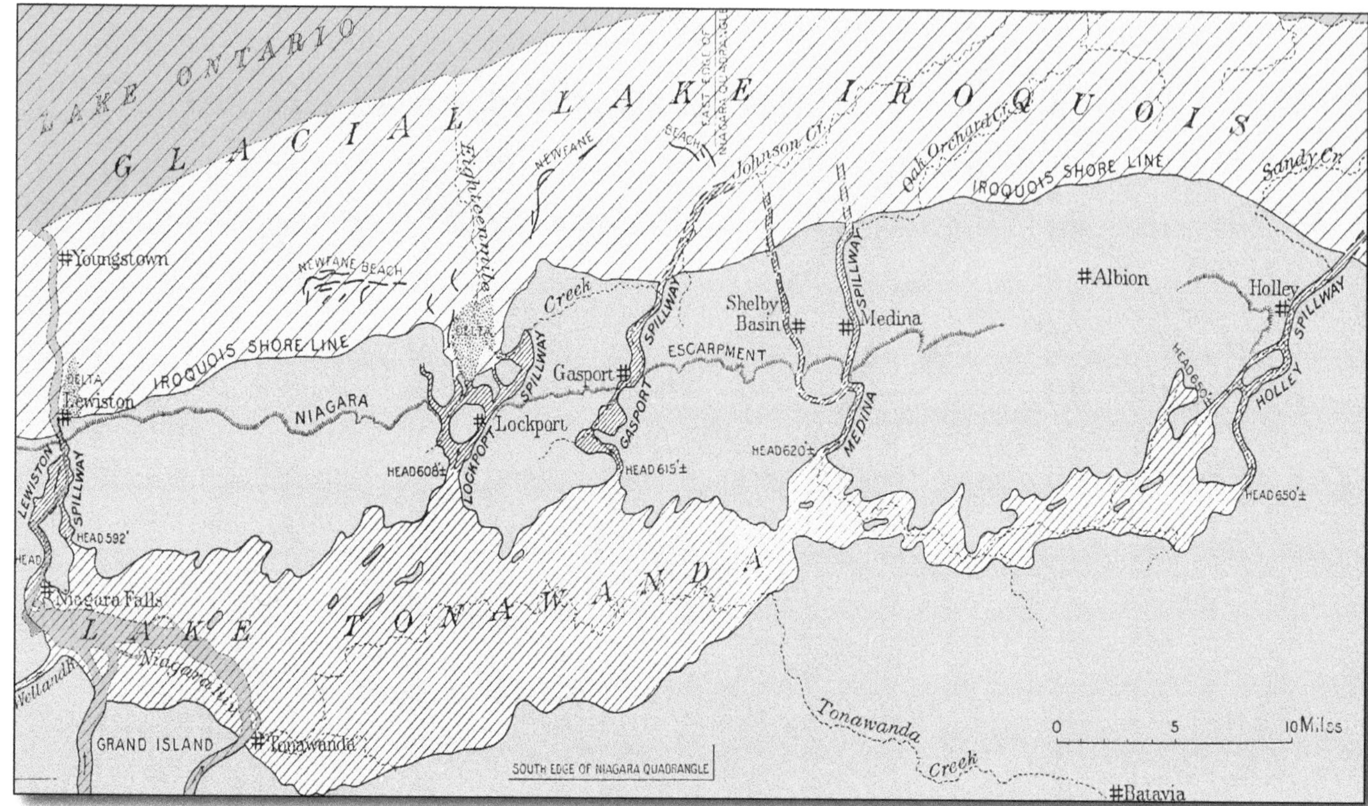

***Diagram 9**: Glacial Lake Tonawanda and the Many Early Falls Over the Niagara Escarpment*

About 8500 years ago, a large, shallow glacial lake called Lake Tonawanda had formed just south of Lake Iroquois due to the meltwater from the retreating ice. Refer to ***Diagram 9**: Glacial Lake Tonawanda and the Many Early Falls Over the Niagara Escarpment.* In addition to the early Niagara Falls, water was spilling over the escarpment at Lockport, Gasport, Medina, and Holley. As the rock layers gradually rebounded back after the weight and pressure of more than a thousand feet of ice were lessened by glacial melting, the Niagara Escarpment tilted at an angle to the west, drastically modifying four of the five falls. This left Niagara Falls as the remaining falls in the immediate area. Examination of the various heads on the diagram show the effects of this rebound. A head is the uppermost part/elevation of a stream. The head at Holley is 630 feet, Medina at 620 feet, Gasport at 613 feet, Lockport at 606 feet and the Lewiston Spillway at 582 feet, with Niagara Falls even lower.

This in combination with the Niagara River eroding through the sill of Lake Tonawanda at Devil's Hole has resulted in its present drainage channel. All that remains of Lake Tonawanda at present is Tonawanda Creek, which serves as the boundary between Erie and Niagara counties which includes a federal wildlife preservation area, the Oakfield-Alabama Swamp.

Comparing the photographs of these other falls, as seen in the samples below, with the spectacular Niagara Falls only dramatizes this rebounding effect.

Photograph 5: *Falls at Holley*

Photograph 6: *Falls at Medina*

Photograph 7: *Royalton Falls at Gasport*

Photograph 8: *Niagara Falls in Comparison*

Niagara Glen

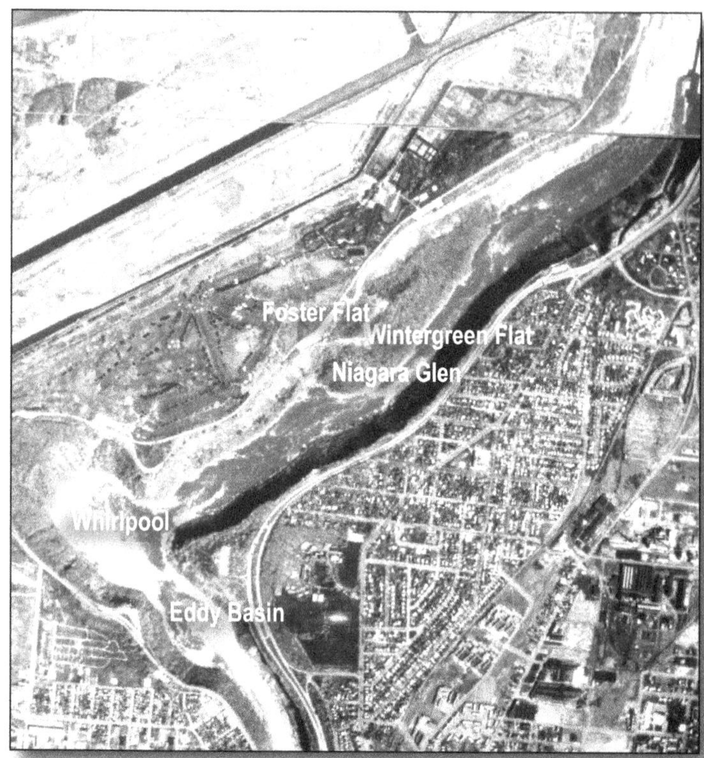

Photograph 8: *Aerial View of Section 3 of the Niagara Gorge*

Photograph 9: *The Dry Falls at Niagara Glen*

Photograph 10: *Rock Fragments in Niagara Glen*

As depicted in ***Photograph 8,*** *Aerial View of Section 3 of the Niagara Gorge,* Section 3 is the longest of the five sections, and includes the largest number and greatest variety of interesting phenomena.

The first prominent feature of the landscape in this section is Niagara Glen, now a Canadian Provincial Park. It represents the remains of an island, similar to the position of Goat Island today, that divided the retreating falls into two uneven parts. With most of the water, and therefore most of its erosive power, concentrated to one side, the main falls quickly left behind a shallow abandoned channel, now called Foster Flat, with its accompanying dry falls, and the rubble from the island at a lower level, now called Wintergreen Flats. See both ***Photograph 9,*** *The Dry Falls at Niagara Glen* and ***Photograph 10:*** *Rock Fragments in Niagara Glen.*

Niagara Glen is one of the most delightful spots along the gorge. It contains many trails and lookouts along the path descending to the river below. The huge pieces of rock that have been undercut by the removal of the softer rock layers below give testimony to the erosive power of water as shown in ***Photograph 10,*** *Rock Fragments in Niagara Glen.*

Photograph 11 A and B: *Potholes in Niagara Glen*

Potholes are in evidence throughout the area. These were formed by the abrasive action of rocks of different hardness caught in the swirling waters at the base of a falls. In this instance, ***Photograph 11 A and B,*** *Potholes in Niagara Glen,* illustrate how harder rock fragments can be carried here by the glacier and were deposited at the base of the Falls. Since they are harder, they grind through the softer local rocks.

These pieces of glacial debris are called glacial erratics or wanderers because they were eroded from distant and dissimilar bedrock by the glacier and transported in the ice to a site far from its origin and deposited on entirely different bedrock as the melting glacier made its retreat to the north.

Photograph 12: *River Level Rapids at Niagara Glen*

Many trails can lead to the river level of the Niagara Gorge and ***Photograph 12:*** *River Level Rapids at Niagara Glen,* start to reveal many of the characteristics of the various rocklayers in the Gorge.

Photograph 13: *Evidence of Crossbedding in a Sandstone Layer* indicates a dry environment during which sand forms different layers and directions as a result of shifting winds and then is transformed into rock by the pressure of the many sediments deposited over time on top.

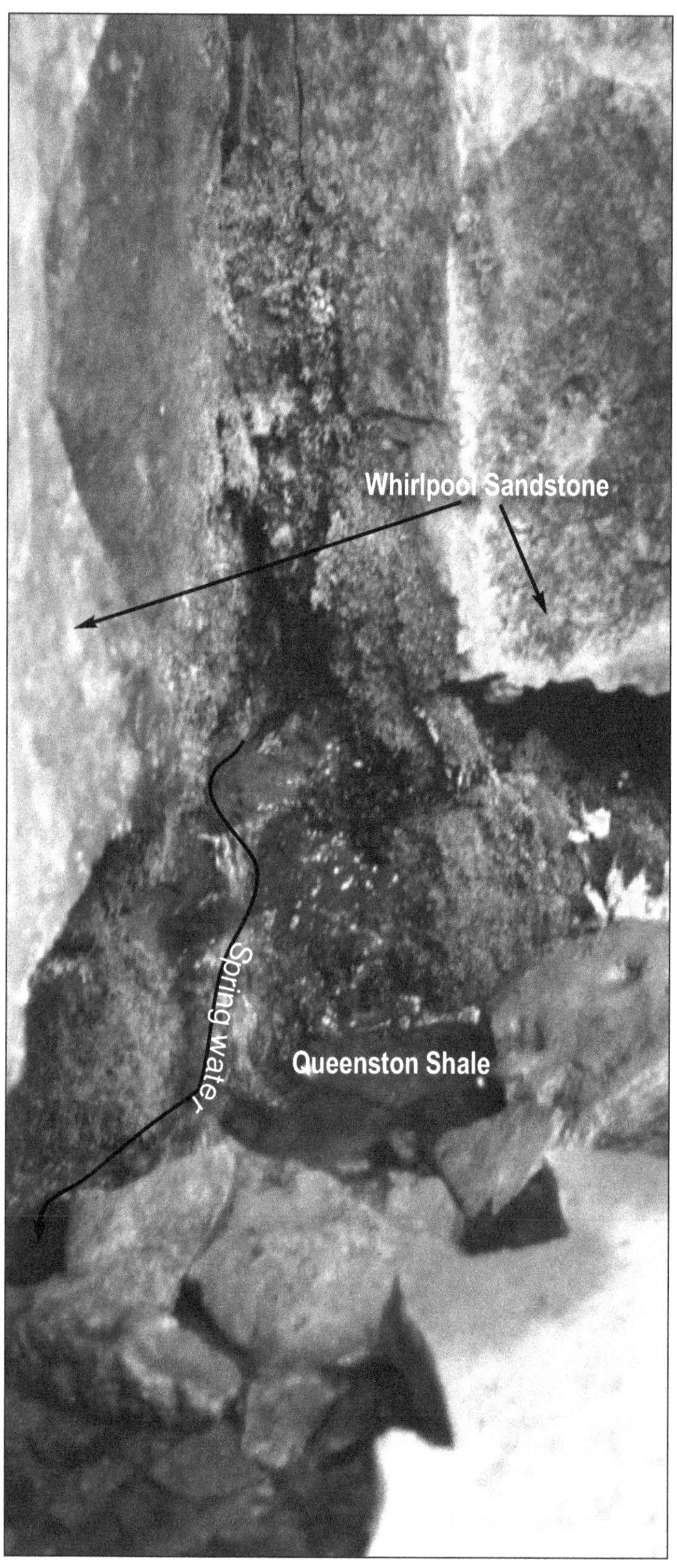

Photograph 14*: Contact Spring* is an example of different permeabilities of the various rocks. Permeability is the ability of rock that allows water to move through it. The lighter colored Whirlpool sandstone on the top is quite permeable as compared to the red Queenston shale shown on the bottom of the picture. It does not allow water to easily pass through it, so where the two layers contact one another, springs appear as the water cannot easily penetrate the shale. These springs are evident all along the gorge where dissimilar permeabilities exist between the various rock layers.

Photograph 14*: Contact Spring*

***Photograph 15**: Along the Trails in Niagara Glen*

The annual temperatures along the trails to the river's edge are lower than above at the top of the gorge as the sunlight is lessened due to the steep slope and abundant trees. This is a haven for native wildflowers, ferns, mosses, ginger, and the seldom seen liverworts that flourish in the cracks and crevices in the huge pieces of limestone and sandstone rubble. See ***Photograph 15**: Along the Trails in Niagara Glen.*

Perhaps the oldest living trees in the Northeastern United States grow on the walls of the gorge. Rare, wild white cedars, some of which are thought to be over 500 years old with 30 foot heights and trunks from 12 inches to 3 feet exist in a tangled mass of branches and surface roots clinging to the soilless, porous bedrocks, their source of a water supply.

Photograph 16 A and B: *The Widening of the Gorge at Niagara Glen*

St. David's River

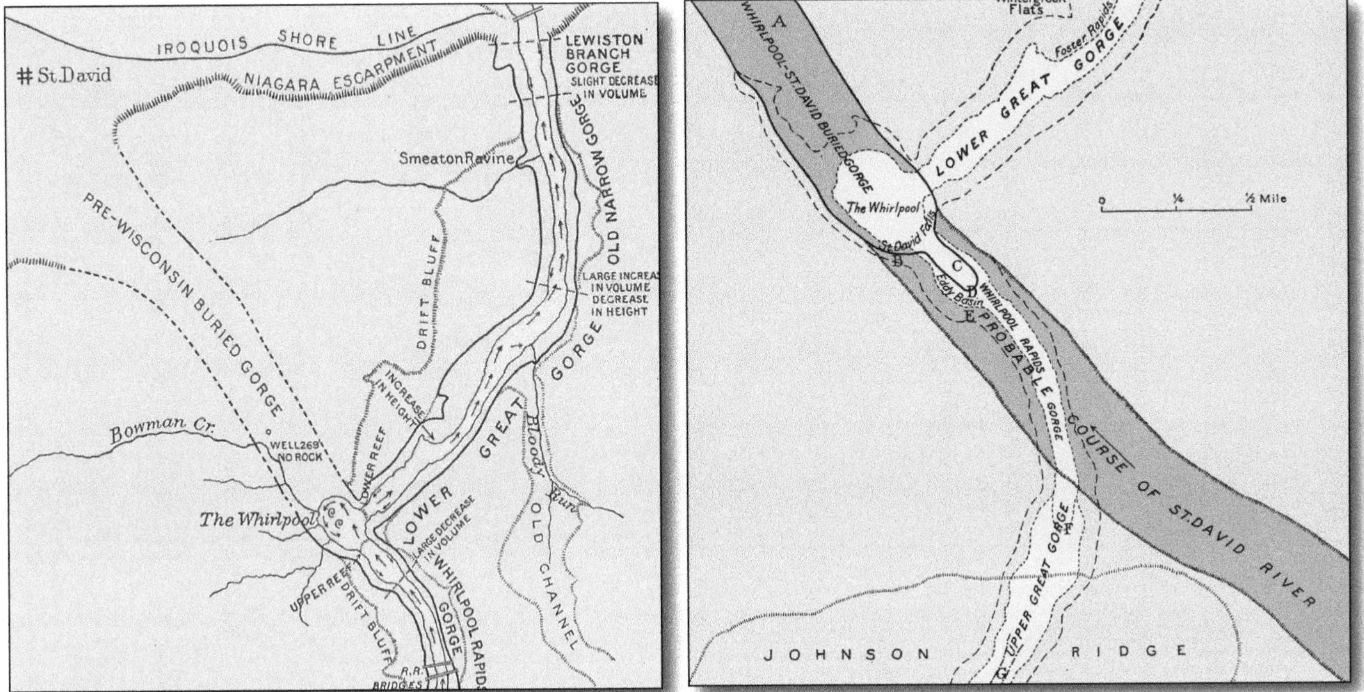

Diagram 10 A and B: *The Buried Pre-Wisconsin St. David's River Channel*

At the southern end of Niagara Glen, the gorge widens and deepens dramatically. This was a result of the opening of the Champlain Outlet mentioned above, followed by a drop in the level of the waters below the falls at Niagara Glen. This drop of the water level consequently raised the height of the falls, increasing its erosive power, and resulted in the widening and deepening of the gorge at this location. Also, the falls once again occupied the full width of the gorge as it passed the island barrier at Niagara Glen. Refer to ***Photograph 16 A and B***: *The Widening of the Gorge at Niagara Glen.*

Farther to the south in Section 3, the gorge makes an unexpected sharp right angle turn as by chance the river uncovered a former river channel, called the St. David River. It had filled with debris deposited by one of the earlier glaciers. Refer to ***Diagram 10 A and B***: *The Buried Pre-Wisconsin St. David's River Channel.* For a time, it was easier for the Niagara River to cut through the loose glacial debris and follow the former river channel than to slowly wear away the solid bedrock that lay directly in its path to the south.

Note the depth of a well 294 feet with the label "no rock" denoting the depth of the glacial debris with still no contact with any bedrock.

Whirlpool Reversal

***Photograph 17**: An Aerial View of the Whirlpool*

As the Niagara River made this dramatic turn, turbulent waters continued to churn and remove glacial debris and carry it away downstream, resulting in a forty-acre pool directly to the west, now known as the Whirlpool of Niagara, as shown in ***Photograph 17**, An Aerial View of the Whirlpool.* The white areas are ice. Note the sharp, right angle turn the Niagara River made in the gorge.

Recognized as unique in all the rivers of the world, the bend in the river is at present occupied by the Whirlpool of Niagara. Water from a now straightened gorge enters the whirlpool from the east and rushes past the northern outlet, consequently establishing a counterclockwise flow of water in the whirlpool. As the water makes this circle, the current is forced to flow under the incoming rush of water that, on the surface, is blocking the northern outlet. This can be seen as turbulent masses of water, boiling up from below as it seeks this northern outlet.

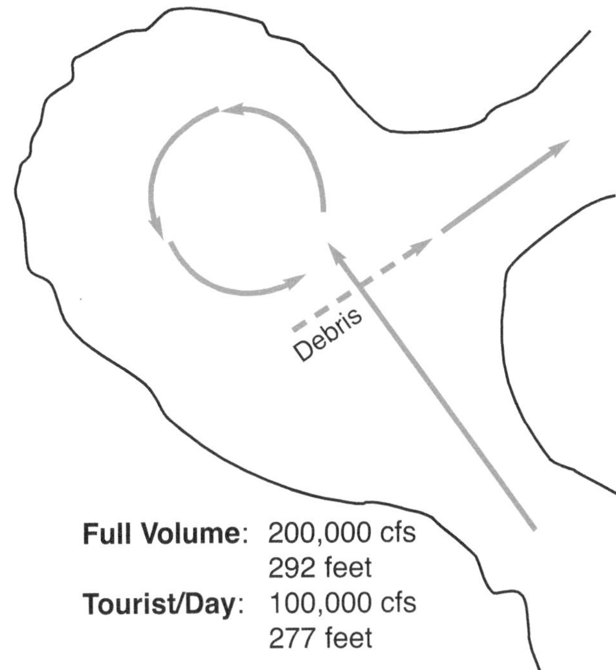

Diagram 11*: Vectors at the Whirlpool*

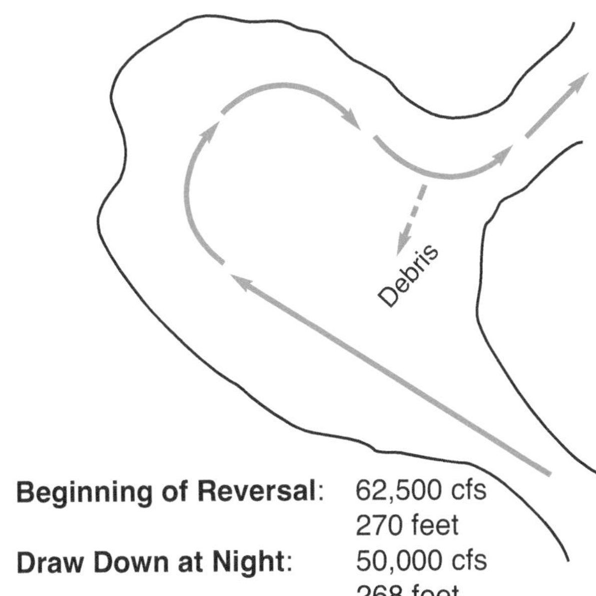

Diagram 12*: Vectors at the Whirlpool*

Normally, without the diversion of water for the two massive power projects that are located downstream from the whirlpool, water flows over the falls at Niagara at the rate of about 200,000 cfs (cubic feet per second). To preserve the beauty of the falls during the day during the tourist season, water flow is reduced to 100,000 cfs, by treaty between the United States and Canada, and to 50,000 cfs during the rest of the year and at night during the tourist season (April 1st to October 1st).

At about 62,000 cfs, the whirlpool reverses from counterclockwise to clockwise. An explanation of this phenomenon seems to be related to the orientation of the old St. David's channel as it empties into the whirlpool.

At reduced flow times, the water follows the slightly angled channel bed and is directed in a more southerly direction, resulting in a clockwise rotation.

At high water levels, the impact on the old channel is minimized by the sheer volume of water flowing in the Niagara Gorge and tends to orient more toward the north, resulting in the counterclockwise flow. This whirlpool reversal is called the Niagara Whirlpool Reversal Phenomenon and is unique in all the natural world. See ***Photograph 17****: An Aerial View of the Whirlpool,* ***Diagram 11 and 12****, Vectors at the Whirlpool*, and ***Photograph 18 and 19****, The Whirlpool at Low/High Water.*

Note the difference between the accumulation of debris/ice in the two photographs. During periods of high volume, the water moves in a counterclockwise direction with the current passing under itself and thereby collecting logs and other debris on the left side of the discharge. During periods of low water (beginning at about 62,000 cfs) the angle of the discharge from the eddy basin is slanted to the left resulting in debris on the right hand side of the discharge flow.

***Photograph 18**: The Whirlpool at Low Water Taken from Whirlpool State Park*

***Photograph 19**: The Whirlpool at High Water Taken from Canada*

Lake Erie Water Levels

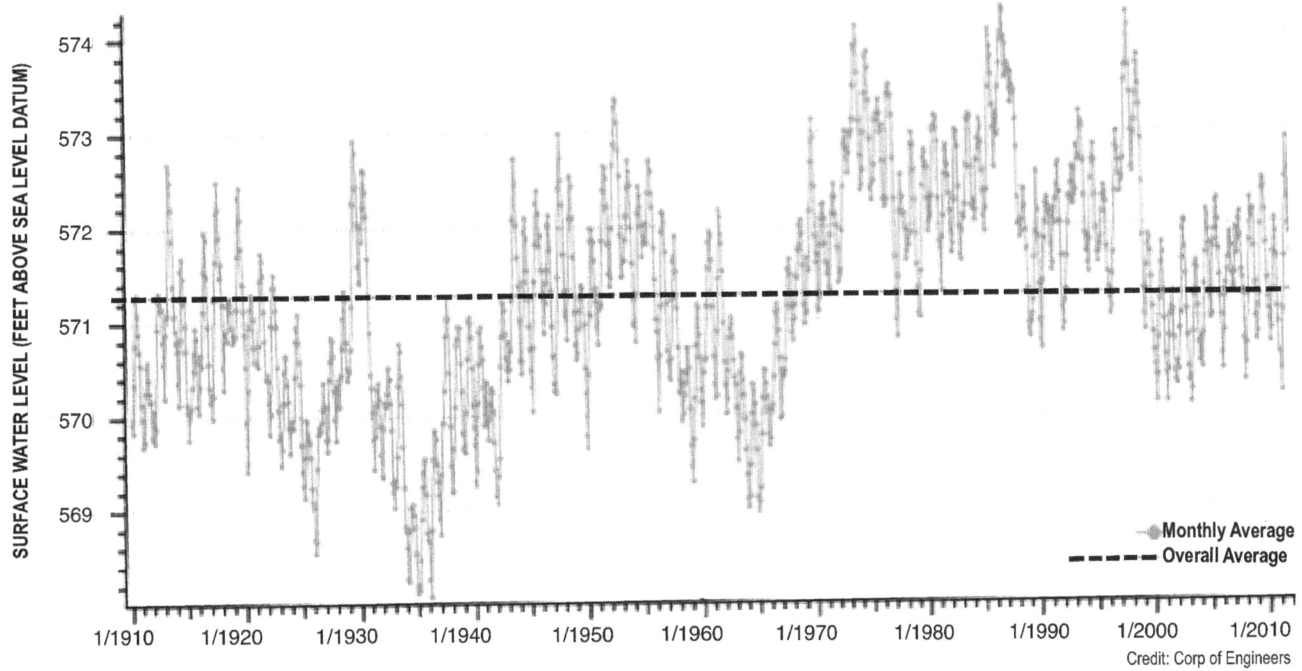

The Niagara River volume and consequently its impact on the Whirlpool water levels, depends on the level of Lake Erie at Buffalo with its 120 year average of 571.4 feet above sea level. Variations occur due to differing annual rain fall in the upper Great Lakes Watershed with seasonal changes peaking in June with lows in January, and short term fluctuations due to wind vectors, storms, and seiches. The flow, before the present conditions of water diversion for power production, translates into 206,000 cfs in the upper Niagara River. If the level of Lake Erie were consistently at 1934 levels, the Whirlpool would naturally be in a constant reversal mode.

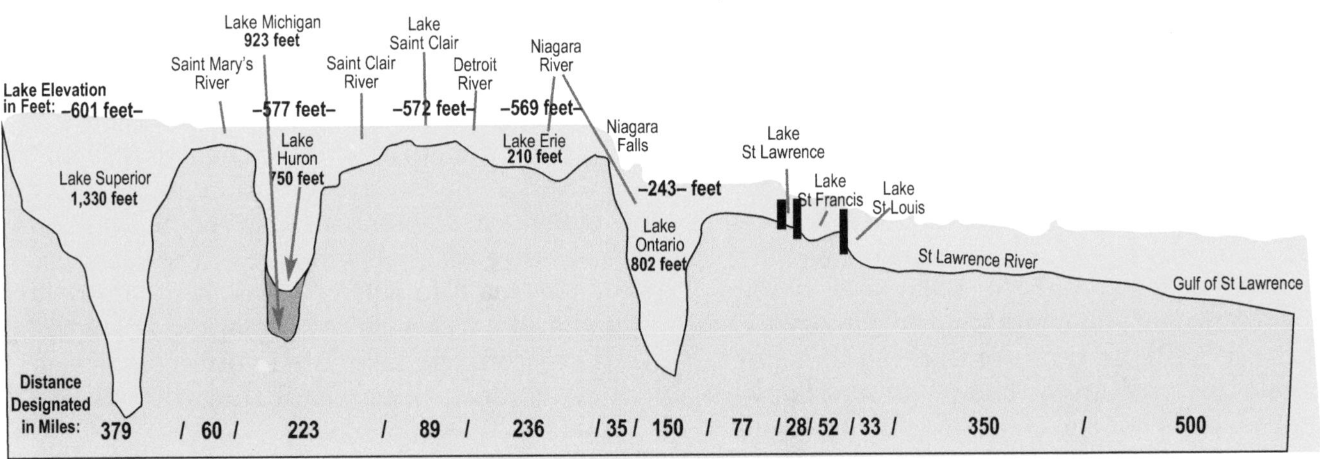

Great Lakes System Profile

Rock Structures

Before proceeding to the explanations of Sections 4 and 5 of the Niagara Gorge, a look at the various rock structures would be appropriate. ***Diagram 13**: Rock Structures at the Whirlpool* corresponds to the following description of the landscape.

***Diagram 13**: Rock Structures at the Whirlpool*

All the rocks in the gorge are sedimentary rocks, that is, sediments were originally laid down under water as sand, mud, silt, shells, or organic or inorganic matter, and gradually compacted into rocks over millions of years and as hundreds of feet of material was deposited on top of them. Each layer has its own characteristics and its own history to tell of the conditions that were prevalent at the time of its deposition. The present rocks were formed south of the equator millions of years ago and through the action of plate tectonics, gradually assumed their present position.

The top layer exposed in the Niagara gorge is Lockport Dolomite, the hard , massive rock that caps the entire area of the gorge. It is basically a limestone formation also containing magnesium carbonate and in some locations along the gorge, as evidence of the warm shallow sea into which it was formed, are large algal and coral reefs. At the origin of the falls at Lewiston, the dolomite is but a few feet thick, while at the present Niagara Falls, a total of 120 feet is exposed.

Also, all the rock layers in the gorge dip to the south at about one-half a degree or fifty feet per mile, so that rock layers that are exposed at Lewiston are no longer visible farther to the south at Niagara Falls.

The second rock layer is the Rochester Shale, a dark gray layer about sixty feet thick and is the reason that the falls has retreated from Lewiston to Niagara Falls over the years. Compared to the Lockport Dolomite, it is easily eroded. This undercutting of the Lockport Dolomite above results in the massive blocks of dolomite seen at the base of the dry falls at Niagara Glen and at the American falls at Niagara.
Rochester Shale is very fossiliferous and contains a large variety of species. It is also rather impervious to water. This, coupled with the fact that it is sandwiched in between two resistant limestone layers, the Lockport Dolomite above and the Irondequoit Limestone layer below, groundwater seeping through the Lockport Dolomite travels along the contact between the two layers and results in the numerous springs visible along the gorge wall. Only about fifteen feet in thickness, the Irondequoit Limestone, because of its hardness, together with the resistant Whirlpool Sandstone a few more layers below, form the three prominent layers visible in the gorge.

Unconformities in the rock layers as indicated in the diagram are periods in the formation of the gorge when there are gaps in the ages of the rocks due to uplift and erosion or non deposition.

The next three layers are quite thin—the Reynales Limestone, a very resistant limestone formation followed by the weak Thorold Sandstone and the greenish shale called the Neahga Shale.

The Grimsby Formation is made up of a mixture red and green shale and sandstone. Its character is variable, with some layers being soft while others are quite durable. Below this formation is the Power Glen Shale a gray, easily eroded layer. As a result, it tends to undercut the Grimsby Formation above in the same way the Rochester Shale undercuts the Lockport Dolomite.

The Whirlpool Sandstone represents a coarse sandstone layer. It can be seen at the waters edge as huge blocks undercut by the soft Queenston Shale below at Niagara Glen and as the flats along the gorge where the water rushes into the whirlpool. It is most visible at low water times. See ***Photographs 20 A and B**: The Whirlpool Rapids at High/Low Water.* This durable layer protects the base of the gorge from erosion and undercutting, resulting in the retention of the narrowness of the gorge.

The last visible layer in the gorge is the Queenston Shale, a mostly red, mixed with green, beds of shale. Because of the tilt of the rock structures in the area, it is quite prominent at the Niagara Escarpment at Lewiston, just above the water level at Niagara Glen and at the waters edge at the Whirlpool. At Niagara Falls it is well below the surface. The three conspicuous layers of Lockport dolostone, Irondequoit limestone and the Whirlpool sandstone clearly show this inclination. (In comparison, the three falls of the Genesee river in Letchworth State Park cascade over three layers as separate falls.).

At the time of the formation of the last part of Section 3, all the water from four of the Great Lakes was going over Niagara Falls making its way into Lake Iroquois (later Lake Ontario), through that part of the St. Lawrence basin not choked by the glacier, then diverted south through the Champlain lowlands to the Hudson River and out to the Atlantic Ocean.

Section 4: Whirlpool Rapids Gorge

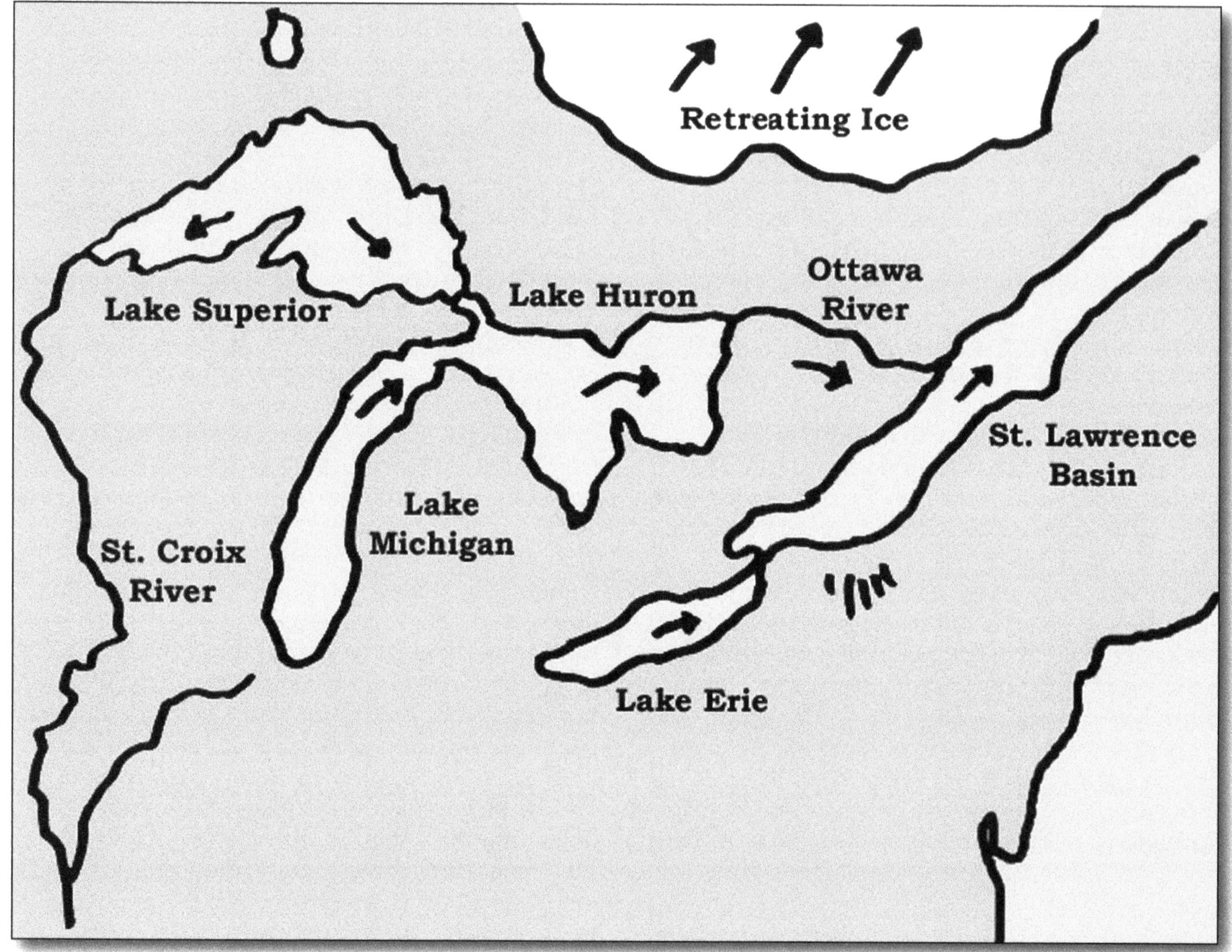

Diagram 14: *Drainage Pattern of the Niagara River in Section 4 of the Gorge*

Over the centuries that followed, dramatic changes took place as the glacier retreated farther north in Ontario, refer to ***Diagram 14***, *Drainage Pattern of the Niagara River in Section 4 of the Gorge.* It uncovered a depressed North Bay and the Ottawa River basin finally revealing the St. Lawrence basin in its entire length including the Gulf of St. Lawrence and emptying into the Atlantic Ocean.

Photographs 20 A and B: *The Whirlpool Rapids at High/Low Water* *High Water*

Low Water

As a result of this lower elevation in the North Bay/Ottawa River basin, water levels dropped significantly in Lakes Michigan and Huron and drained in this new path to the now open St. Lawrence basin. Seawater from the Atlantic also flowed into the depressed basin, leaving Lake Iroquois in a brackish condition. The Champlain Outlet, being higher in elevation, ceased to be part of the main drainage pattern. Lake Superior's water continued at its former depth, held back by a rock sill at its outlet.

At this time in the Niagara Gorge, with only the water from Lake Erie flowing over the Falls, the erosive power of the Falls, now located at the head of the Whirlpool Eddy basin, was greatly reduced to about 15%. No longer was the river able to gouge out a deep channel in the river, let alone a plunge pool at the base of the Falls. As a result, a long series of rapids developed, now known at the Whirlpool Rapids. See ***Photographs 20 A and B****: The Whirlpool Rapids at High/Low Water.*

Section 5: Upper Great Gorge

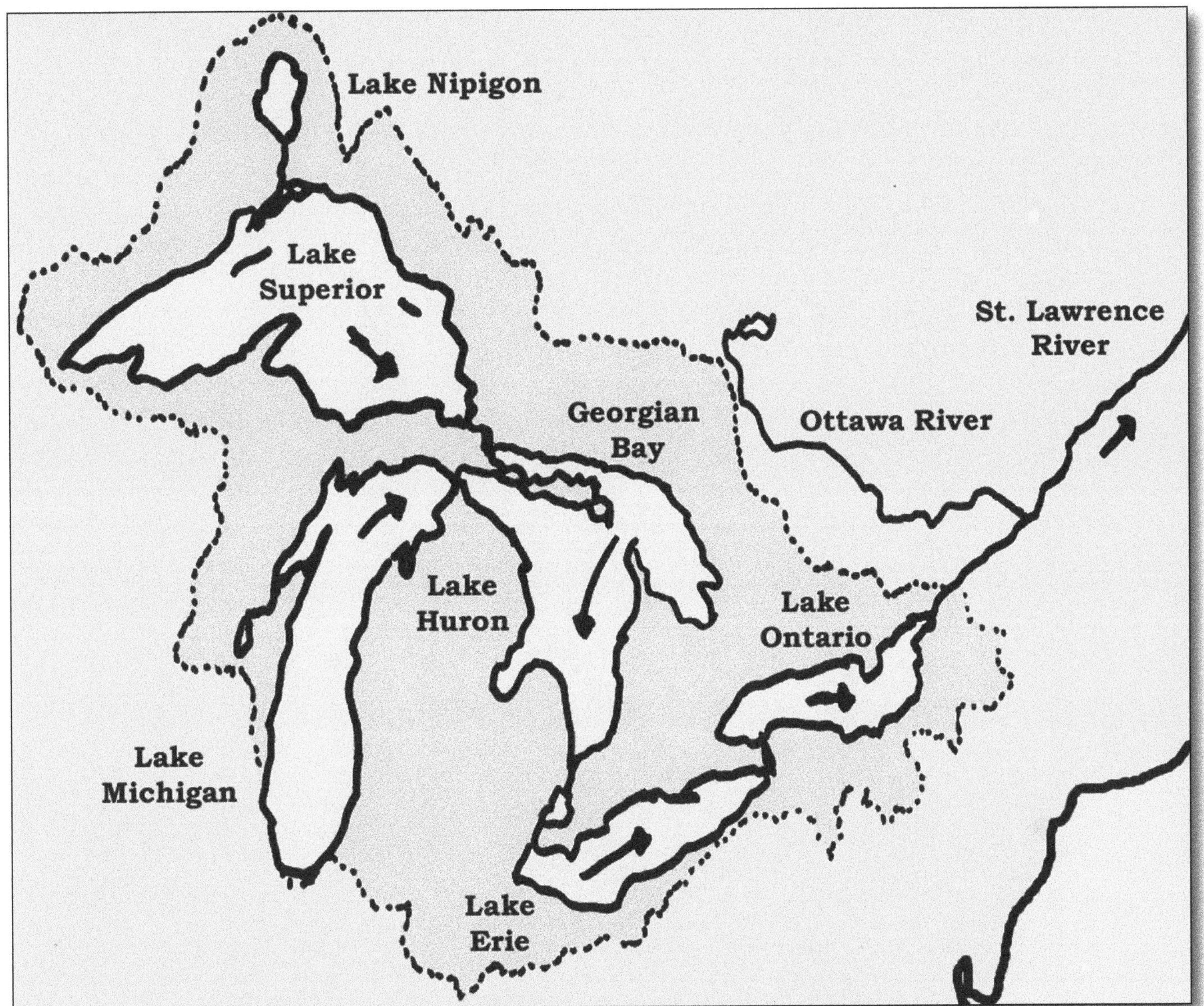

Diagram 16: *Rise of the North Bay/Ottawa River Basin and the Beginning of the Modern Drainage Pattern of the Five Great Lakes*

Entering Section 5, the gorge becomes much wider and deeper, and assumes its modern shape and present drainage pattern. Over the hundreds of years since the Ottawa River path, the rock layers have rebounded upward, forcing the water to drain via the lowest elevation levels, cutting through the sill of glacial Lake Tonawanda and assuming its modern pathway. At present, all four Great Lakes flow over the Falls at Niagara into Lake Ontario, to the St. Lawrence River and then to the Atlantic Ocean. See ***Diagram 16***, Section 5, *Rise of the North Bay/Ottawa River Basin and the Assumption of the Modern Drainage Pattern of the Five Great Lakes*, and ***Photograph 21,*** *Aerial View of Section 5 with White Rapids and Snow Covered Areas and* ***Photograph 22***, *The Modern Niagara River Gorge/Sill of Lake Tonawanda.*

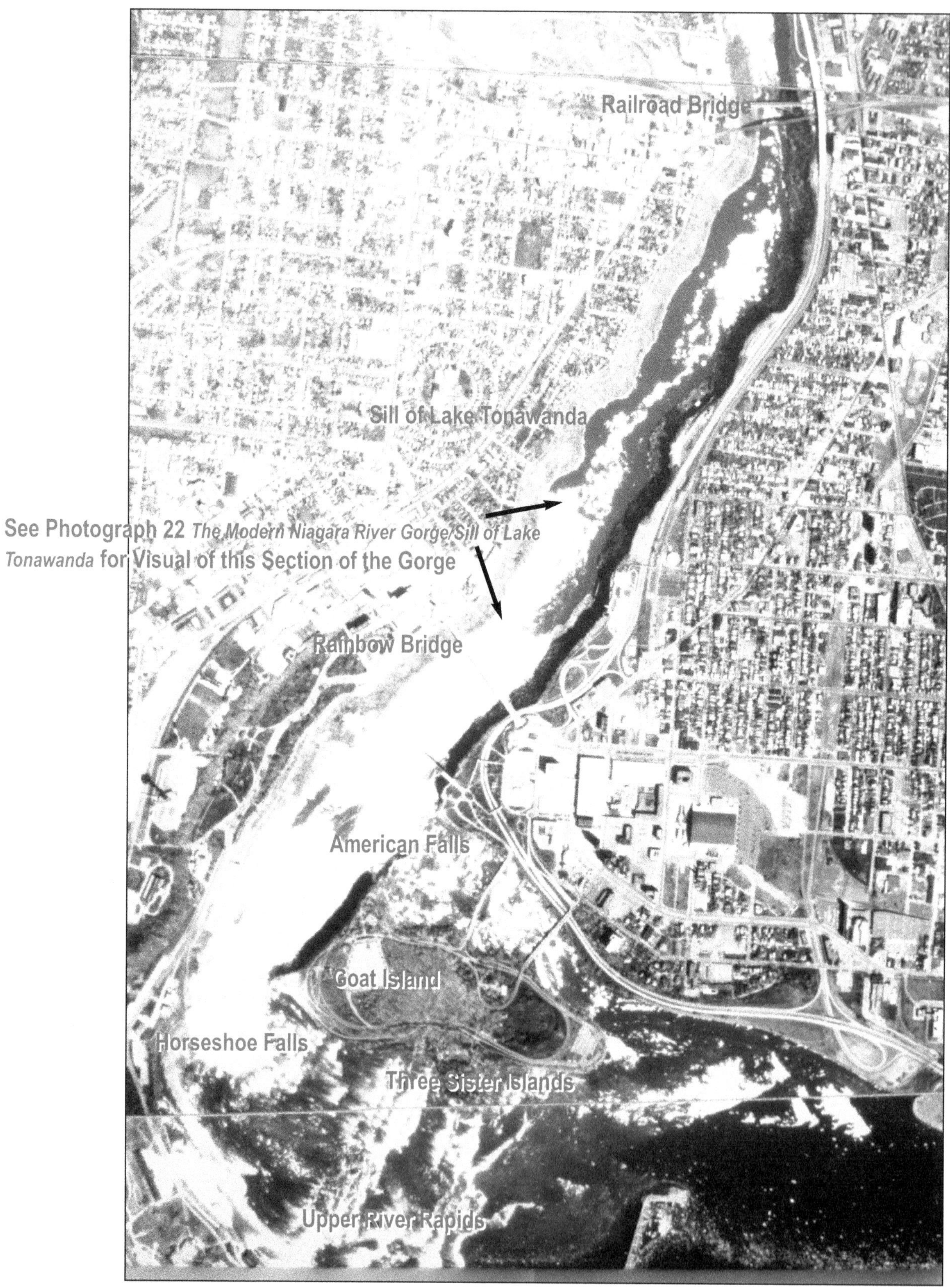

Photograph 21: *Aerial View of Section 5 with White Rapids and Snow Covered Areas*

Photograph 22: *The Modern Niagara River Gorge/ Sill of Lake Tonawanda*

Photograph 23: *Maid of the Mist Pool*

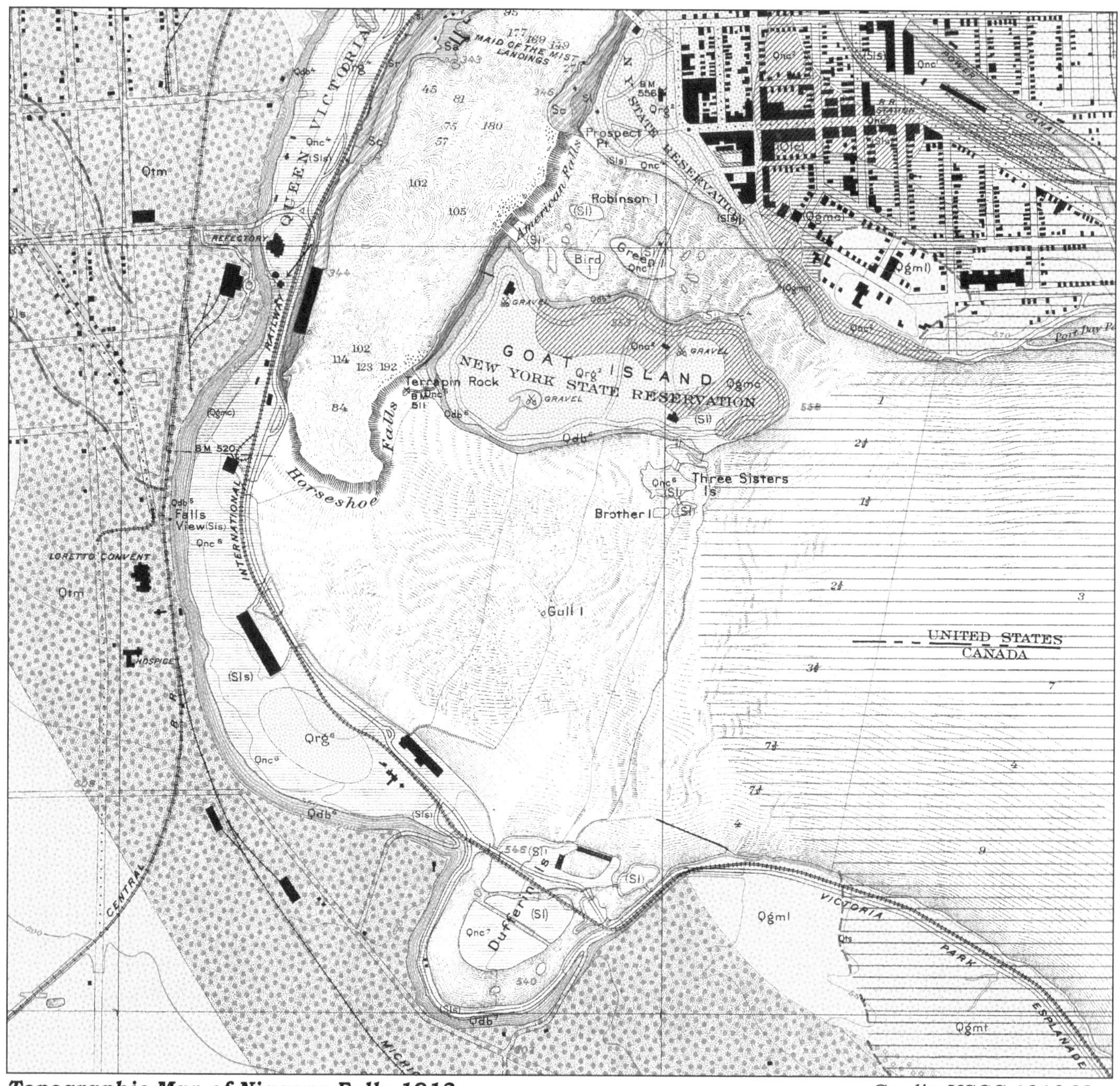

Topographic Map of Niagara Falls 1913

Credit: USGS 1913 Map

The previous three images of Section 5 provide a multi-dimensional composite of the gorge. The aerial view with ***Photograph 21****: Aerial View of Section 5 with White Rapids and Snow Covered Areas,* represents the final section of the gorge and Niagara River on its journey to Lake Ontario. The ***Topographic Map of Niagara Falls from 1913*** represents elevations, depths and rock structures at recent Niagara Falls. Both ***Photograph 22****: The Modern Niagara River Gorge/ Sill of Lake Tonawanda* and ***Photograph 23****: Maid of the Mist Pool* represent in a photographic image of the current Niagara River and Gorge that tourists, hikers and locals would view as a popular area for hiking.

Thundering Niagara Falls

Between six and seven hundred years ago, the Horseshoe Falls, the original Niagara Falls, separated from the American and Bridal Veil Falls and today represents about 90% of the water flowing over the three falls. In the past, the rate of retreat of the Falls has averaged between three and eight feet a year, depending on the amount of water and the height of the Falls. Since the diversion of water for the two power projects, the rate of retreat has been reduced to about a foot a year.

The American Falls has the length along its crest of about 800 feet, the Bridal Veil Falls about 100 feet, and the Horseshoe Falls about 2500 feet.

The height of the Falls vary with the season, but are generally between 170 to 185 feet, with the plunge pool at the base of the Falls of approximately equal depth.
See ***Photographs 24*** *to* ***28***.

Photograph 24: *The American and Bridal Veil Falls*

Photograph 25: *The Crest of the American Falls*

Photograph 26: *The Bridal Veil Falls*

Photograph 27: *Horseshoe Falls*

Photograph 28: *Brink of the Horseshoe Falls*

Photograph 29: *American Falls Rapids*

Photograph 30: *American Falls Rapids*

Photograph 31: *Rapids in winter*

The De-watering of the Niagara Falls

From June to November 1969, both the American and Bridal Veil Falls were de-watered to determine if the erosion of the falls could be controlled. This unique opportunity to view the underlying rock structure of the Falls is seen in ***Photographs 32*** *to* ***35.***

Photograph 32 A and B*: Coffer Dam Construction Across the Channel of the American Falls*

***Photograph 33 A**: The De-watered Rapids Above the American Falls*

***Photograph 33 B**: The Rapids Above the American Falls at Full Volume*

Photograph 34 A: *The De-watered Crest of the American Falls*

Photograph 34 B: *The Crest of the American Falls at Full Volume*

***Photograph 35 A**: The De-watered American and Bridal Veil Falls*

***Photograph 35 B**: The American and Bridal Veil Falls at full Volume*

Ice Bridge/Winter

The barren rock of the dry falls in summer, contrasted with the cataract in winter, which may form an ice bridge beneath the American Falls, thereby joining the two countries with a treacherously rugged landscape, providing interest and variety. During more misty and windy times, trees, shrubs and rock structures along the top of the gorge become coated with ice and may present a spectacular or errie scene.

See ***Photographs 36 – 39***: *Ice Formations Along Niagara Falls and Gorge* below.

36

37

38

39

Photograph 40: *Winter Mounds on the American Falls*

Summaries

The rock layers and water levels of the events in Niagara Gorge's story can be summarized as shown in ***Diagram 16***, *Summary of the Rock Layers and Water Levels by Sections.*

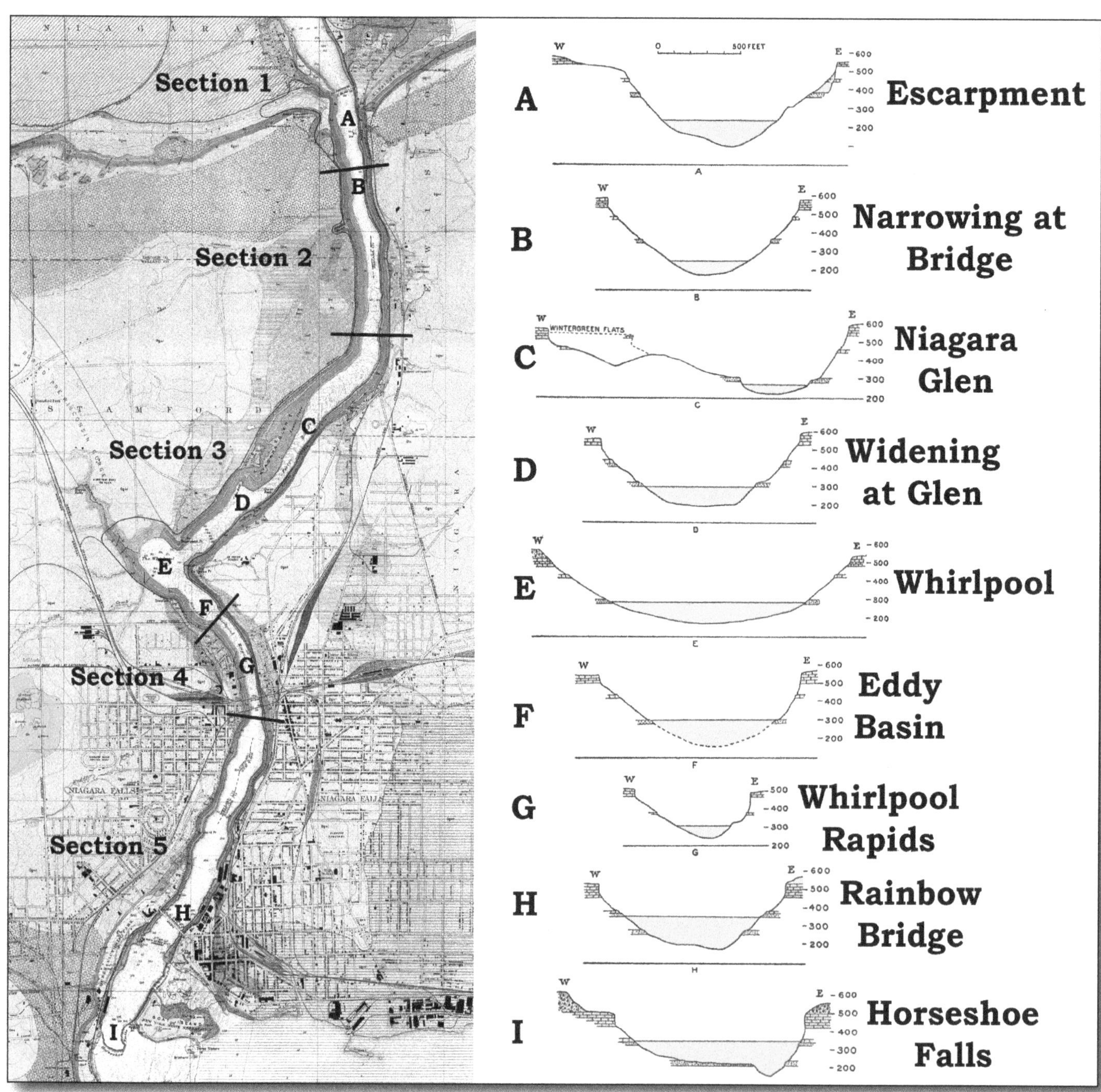

Diagram 16: *Summary of the Rock Layers and Water Levels by Sections*

***Diagram 14**: The Overall Summary of the 5 Sections of the Niagara Gorge* below indicates the varying changes in water volume, height of the falls and resulting width and depth of the gorge.

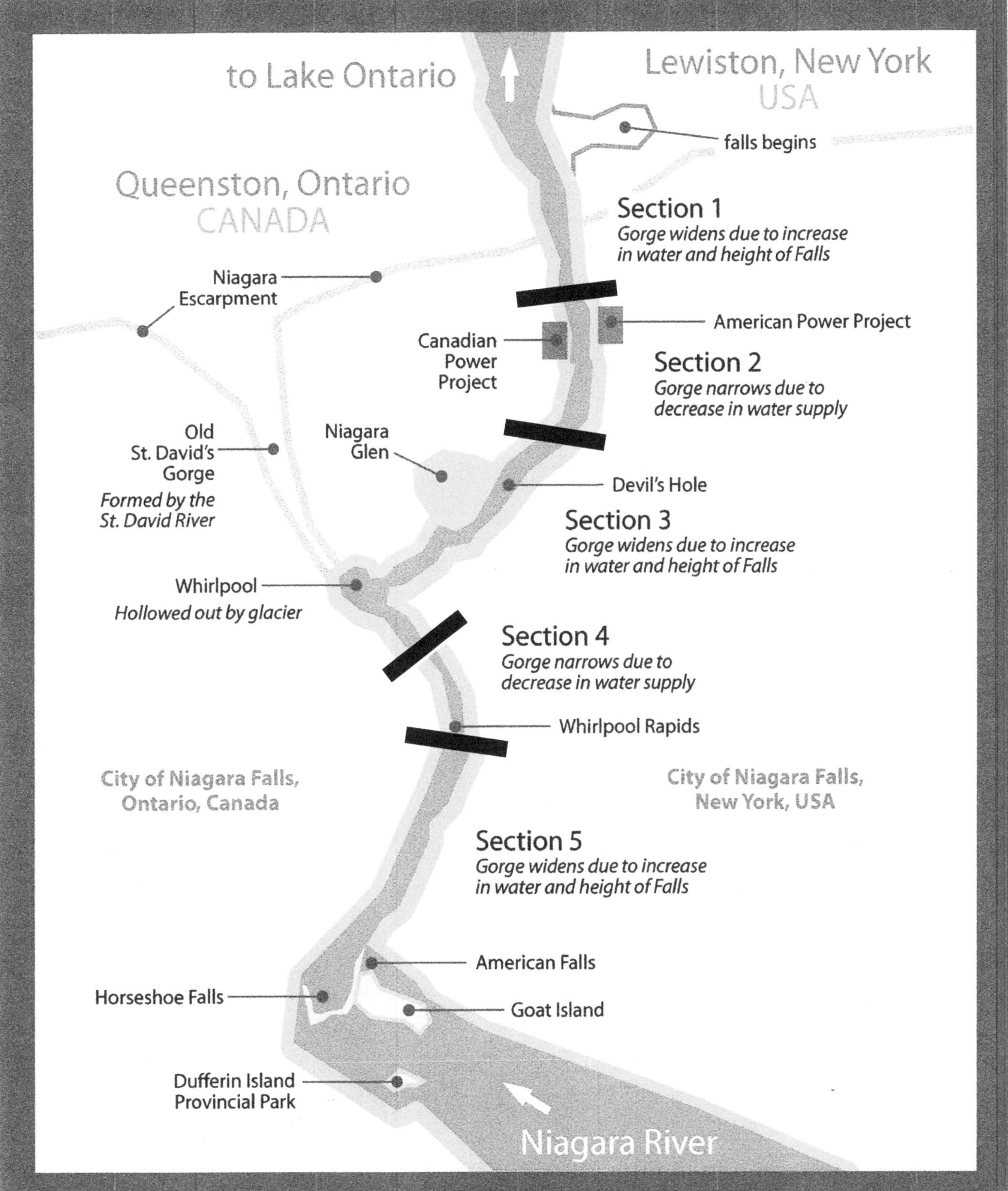

In the future, it is predicted that the falls will erode to the Upper Cascades with the height of the single falls at about 230 feet, then will slowly lower as uplift continues and dry up and resume a drainage pattern as in the pre-Niagara times.

This depiction of all five sections presents a coherent summary of all the particular events that has made the Gorge what it is today—a marvel of the various natural forces that, over time, either singularly or in combination, have produced a majestic and awesome panorama of sights, sounds, and emotions for ages to come!

***Photograph 41**: Brink of the Horseshoe Falls*

***Photograph 42**: Sunset at Niagara Falls*

Glossary

Alluvium Sediment deposited by a stream.

Caprock A layer of resistant rock that caps the top of a landform, protecting the underlying rocks from erosion.

Contact Spring A flow of groundwater that emerges naturally at the boundary between rock surfaces of differing permeabilities

Continental Glaciation An enormous mass of flowing ice that covers a significant part of a large landmass.

Crossbedding Layering of material resulting from deposition due to differing directions of the erosional agent.

Cuesta A landform that slopes gently back from a steep cliff, in this case, the Niagara Escarpment.

Delta A low level plain resulting from stream deposition into a rather still body of water.

Drift Sediment directly or indirectly deposited by a glacier.

Drumlin A streamlined, asymmetrical landform that is created when a glacier deforms previously deposited till with the steep side of the hill facing the direction from which the glacier advanced and with long tapering sides.

Esker A winding ridge of sediment formed by a stream that flows beneath a glacier.

Glacier A slow moving mass of dense ice that flows under its own weight.

Glacial Abrasion An erosional process caused by the grinding action of a glacier on rock.

Glacial Erratic Boulders that have been plucked and transported a great distance from their orignal location before being deposited.

Glacial Grooves Deep furrows in the bedrock produced by glacial abrasion.

Glacial Outwash Sediment deposited by meltwater streams from a glacier.

Glacial Plucking An erosional process by which rocks are pulled out of the ground.

Glacial Rebound A springing back of rock structures after the weight of the glacier is lifted.

Ground Moraine Unassorted glacial till that forms when the retreat of a glacier is steady and slow, resulting in an irregular depositional pattern.

Glacial Till Sediment deposited directly by the glacier.

Head The upper most part/elevation of a stream.

Hydrologic Cycle/Water Cycle A model that illustrates the way that the various forms of water are stored and moved from one reservoir to another.

Kame A large mound of sediment deposited along the front of a slowly melting or stationary glacier.

Kettle Hole A depression in glacial drift formed when a block of ice breaks off the glacial front, is buried and then melts.

Kettle Hole Lake A kettle hole that is transsected by the water table.

Marginal Lake A lake formed at the glacial front.

Outwash Plain An extensive, low relief landscape caused by the deposition of glacial outwash.

Pothole A hole made by the swirling and grinding action of a glacial erratic on softer, native rock at the base of a waterfall.

Process A series of events that can be measured resulting in possible predictable outcomes.

Recessional Moraine Ground moraine deposited in ridges as a result of irregular melting of a retreating glacier.

Seiche Long waves that move back and forth as they reflect off the opposite side of a basin.

Spillway A passageway through which surplus water escapes.

Terminal Moraine Ground moraine deposited as a ridge at farthest extent of a glacier.

Weir A natural dam in a stream.

Eurypterid:
The Official New York State Fossil

Acknowledgements

The following have contributed their time and talent in the production of this publication:

- Mrs. Veronica A. Young: Layout and design suggestions, patient follow through with their application and revisions.
- Mr. Marc A. Young: Technological, diagram skills and computer expertise.
- Ms. Mary Jean Syrek: Editing skills.
- Dr. Mark Donnelly: Guidance during the publishing process. Layout and design expertise

Credits

Geological Atlas of the United States
Niagara Folio, New York
by E. M. Kindle and Frank B. Taylor
Washington, D.C. 1913
Engraved and printed by the U.S. Geological Survey Folio #190

Aerial Photography by Aerial Images
Tonawanda, New York
Copyrighted Photos Used With Permission

Photographs
Dr. Paul A. Young and Dr. Mark Donnelly

Research Paper
Niagara Whirlpool Reversal Phenomenon
Robert B. MacMullen PE, private paper, 1973

Diagram 17, *The Overall Summary of the 5 Sections of the Niagara Gorge*
John Arnold

About the Author

Dr. Paul A.Young is a Professor Emeritus/Science Education, Education Division Outstanding Educator, and a Koesseler Distinguished Professor at Canisius College in Buffalo, New York.

He was born in Western New York and is an alumni of Buffalo State College, Canisius College and the University of Georgia. As a navy veteran, his tour of duty took him to many Pacific Islands and later through his travels, gave him the opportunity to expand his interests in global science while developing a deeper appreciation for his local environment.

Dr. Young has had a lifelong interest in the natural sciences and science education. For years he has researched, viewed and photographed many facets of Niagara Falls and its gorge. This focus has culminated in this informative, educational and interesting publication.

In addition to his teaching, Dr. Young has led many natural science field trips and tours for students, professional staff and community groups to the Falls. He continues to work as a speaker for many organizational groups on a variety of topics and issues in science education.

CPSIA information can be obtained
at www.ICGtesting.com
Printed in the USA
LVHW070850060322
712750LV00010B/422